# Edible Plants

## of Newfoundland and Labrador

Peter J. Scott

**BOULDER
PUBLICATIONS**

Library and Archives Canada Cataloguing in Publication

© 2010 Scott, Peter J.

Edible Plants of Newfoundland and Labrador / Peter J. Scott.

Includes bibliographical references and index.

ISBN 978-0-9809144-9-8

1. Plants, Edible--Newfoundland and Labrador. I. Title.

QK98.5.C3S35 2010      581.6'3209718      C2010-900694-1

Published by Boulder Publications

Portugal Cove-St. Philip's, Newfoundland and Labrador

www.boulderpublications.ca

Design and layout by Vanessa Stockley, www.GraniteStudios.ca

Front cover photo by Todd Boland

Printed in China

We acknowledge the financial support of the Government
of Newfoundland and Labrador through the Department
of Tourism, Culture and Recreation.

We acknowledge financial support for our publishing program by the
Government of Canada and the Department of Canadian Heritage through the
Canada Book Fund.

# Contents

# Introduction

There is a long tradition in Atlantic Canada of eating edible plants found in the wild. Everyone is familiar with the variety of berries found in the countryside, a number of which are produced in abundance. For many people berry-picking is often a time to reflect and commune with nature. In the past, for many families the harvest supplied the table throughout the year and paid for school supplies. In addition to berries, some herbs were also collected and eaten.

Traditionally, people grew root crops, caught game, picked berries, and lived well. Fresh produce was only available if it was grown or harvested locally. Professor M.L. Fernald of Harvard University, who visited Newfoundland and Labrador each summer for years in the 1920s, related the following:

> Entertained at one of the great cod-fishing "rooms" on the coast of Labrador, where vegetables are the products of tin-cans, a party of botanists, delighted with the thickets of *Atriplex gladriuscula* [Smooth Orach] growing 6 feet high and freely branching on the refuse from the fish-cleaning sheds, brought in a large bundle of leafy tips and requested to have them cooked. After the meal the genial host, native of southern Newfoundland, said to his daughter, "This is a godsend. Now we can induce Mother to come down to the Labrador if she can have fresh vegetables." (Fernald and Kinsey, 1943, p. 182)

There are many reasons why people may want to embark on a culinary adventure. Some may want to re-establish ties with the natural world and collecting forays are part of the pleasure of using wild plants. Wild foods often have a distinctive flavour and smell. Also, these wild plants of nature produce on a renewable and sustainable basis.

Eating wild plants also benefits one's health. Herbs are rich in chlorophyll, vitamins, and minerals. Health-promoting supplements are extracted from some of our plants; for example, resveratrol, the key ingredient in red wine, is often extracted from Japanese Knotweed for supplement tablets. Berries are often rich in antioxidants. There is a long history of medicinal uses of wild plants by mankind, and home remedies were used in communities throughout Nova Scotia, New Brunswick, Prince Edward Island, and Newfoundland and Labrador. Medicinal uses are beyond the scope of this work but here is a point to be considered: it has been noted that when substances are isolated from a plant, they often lose their efficacy. Enjoying freshly collected berries and herbs may be better than swallowing a tablet.

My earlier book, *Some Edible Fruit and Herbs of Newfoundland* (1979), presented only those plants that are common, found in relative abundance, and tasty. Today there is a renewed interest in using plants that are available locally to supplement the daily diet. For this present volume many sources were consulted and those species that are considered edible are included. They are grouped by the habitats in which they are found. Those which could appeal to most are illustrated and described in some detail.

In addition to being harvested for the table, these plants are also used in such complementary activities as perfume- and soap-making as well as numerous crafts.

**Using this book:**

* Not all foods agree with everyone. For example, strawberries may evoke thoughts of romantic breakfasts for some, but itching and hives for others. Try new foods in small portions. I have eaten all but a few of the plants in this book. Several I was not able to harvest at the right time, in particular some species found on the west coast of Newfoundland. Finding a bottle of spruce beer to enjoy continues to elude me.

* Collect with respect. Check the abundance of a plant in an area before harvesting so that its population is not destroyed.

* Before collecting and eating any plant, make sure of its identity. If in doubt, find out!

* Include wild foods in their season in your diet. Some appeal to many folk, while others are an acquired taste.

* Collecting young greens in the spring may involve finding mature plants

in the autumn and marking the location. Remnants of the previous year's growth will often pinpoint the exact spot in the spring.

\* Harvesting excursions for wild food can take several forms. A few plants for a meal can be picked on a ramble. A family outing may be organized, especially for berry-picking. City-dwellers may collect in vacant lots and fields. Growing weeds, on purpose, may present a philosophical challenge for many, but here are some suggestions: Allow weeds to mature a little before pulling them in the spring. They will then be just right for the pot. A corner can be used to produce for the table. The seeds of several plants are available commercially – Garden Sorrel, Winter Cress, and Dandelion. I have grown these in rows; the Dandelions were magnificent.

\* For some of the species in this book, the description mentions that there are several species but these are not described. This is done in cases where they closely resemble each other and are all edible. The Latin names and common names of the species are listed in the Appendix. Glen Ryan's *Native Trees and Shrubs of Newfoundland and Labrador* (1978) can be used if you wish to identify the species of a woody plant. Another excellent source is John Maunder's website: www.digitalnaturalhistory.com/flora.htm. The family name you will need to know in order to use the website can be found in the Appendix of this book.

*1*

# HEATHS

*Areas of short (15-30 centimetres tall)*
*shrubs and herbs interspersed with rocks,*
*heaths are found on windswept headlands and hills.*
*The plants are kept short by their exposure to the wind.*

# Partridgeberry/Redberry/ Mountain Cranberry/Lingonberry

*Vaccinium vitis-idaea* L.

*"The Beothuks dried berries with oil from sea mammals such as seals to make pemmican-like food."*

(Obed, 2008)

In many parts of North America the Partridgeberry of Newfoundland and Redberry of Labrador are known as Mountain Cranberry or Foxberry; in Scandinavia this berry is known as Lingonberry.

This attractive plant is a circumboreal species found throughout eastern Canada. It grows flat on the ground on headlands but is about 10 centimetres high in more sheltered sites. It can be identified by its glossy, oval leaves that have a crease at mid-vein. The fruit (about 1 centimetre in diameter) has a depression at its apex surrounded by small flaps of the sepals. It grows in most habitats but picking is best on barren hills and heaths.

The fruit are home to the fruitworm which is the larva of the moth *Grapholita libertina*. This fruitworm bores through the fruit and emerges in late August or early September, after which the fruit can be picked. The date of emergence varies, so check a few berries for fruitworm (white in colour, 3-4 millimetres long) before you start picking.

Buckets are the usual vessel used for picking berries but people in Newfoundland have traditionally picked partridgeberries in flour sacks or pillowcases. The flavour of the berries is enhanced by a frost; they are then too juicy for harvesting but make a pleasant snack on an autumn hike. The berries were traditionally stored in barrels of water in a cellar or on a porch but now they are stored in the freezer.

The tart berries make jam, sauces, wines, pies, tarts, and quick breads. A small bowl of jam served with thick cream is a quick and delicious dessert.

# Crowberry

*Empetrum nigrum* L.

*"They are sometimes used by the residents of this island [Sable Island] in the manufacture of a slightly alcoholic drink. The berries are crushed, then after the addition of sugar or molasses the juice is put in a dark air-tight receptacle until fermentation takes place."*

(Fernald and Kinsey, 1943)

Each autumn migrating shorebirds visit headlands and feast on the plentiful crowberries.

The plant forms dense mats about 10-15 centimetres thick called Blackberry Heath. The rod-shaped leaves are densely crowded on the stems, and on its undersurface each leaf has a whitish line along its length. Crowberries are one of the first plants to bloom in spring, often in mid-March. Small reddish male and female flowers are borne on separate plants. The fruit, which start to ripen in late July, are found in the woods, on barrens, and on headlands. It is a circumboreal species found throughout the Maritimes and Newfoundland and Labrador.

There are two other species:

Pink Crowberry (*Empetrum eamesii*), found in Cape Breton, eastern Quebec, southeastern Labrador, and in much of Newfoundland.

Purple Crowberry (*Empetrum atropurpureum*), found in eastern Quebec, Nova Scotia, New Brunswick, Prince Edward Island, western and northeastern parts of Newfoundland, and southeastern Labrador.

The berries of both species are edible.

Picking these berries is pleasant as you can recline on a carpet of plants on an early autumn day. Crowberries are juicy but have little flavour. They make a good snack while on a hike and can be used to make wine or crowberry pudding. Store them in the freezer.

# Angelica

*Angelica atropurpurea* L.

*"In an Island of the North, called Island [Iceland], where it groweth very high, it is eaten of the inhabitants, the barke being pilled off."*

(Gerard, 1633)

This member of the Parsley Family can grow quite tall – up to 1 metre or more. Its stem is smooth, unlike the hairy Cow Parsnip (*Heracleum lanatum*), and has compound leaves. The flowers are white and borne in large (about 15 centimetres in diameter) umbels. Angelica grows in small, sheltered depressions around the coastline; it is found in the Maritimes, central and western Newfoundland, Labrador north to Makkovik, and west to James Bay.

The young flowering stems and leaf stalks are collected in the spring, as they become bitter during the summer. These are used as a vegetable or are candied (see Recipes). Peel the young stems as they have fibres like the related celery. Chop them finely for a salad or boil for three minutes and serve with butter, salt, and pepper. If the taste is too strong, boil in two waters.

Essential oils distilled from the seeds and roots of Angelica are used in perfumes and as flavourings in gin and vermouth and liqueurs such as Chartreuse.

# Scurvygrass

*Cochlearia officinalis* L.

*"But if a captain stowed a supply of scurvy grass, as the 18th-century English explorer Capt. James Cook did, the sailors were safe [from scurvy], for the herb is rich in that vital nutrient [vitamin C]."*

(Gordon et al., 1986)

Sailors and early settlers, who often suffered from scurvy, ate this plant, which is an excellent source of vitamin C. It is not a grass but a member of the Mustard Family.

Scurvygrass sprawls flat on the ground in windy sites but is about 15 centimetres tall when sheltered. This plant has rosettes of broad heart-shaped leaves, white flowers, and globe-shaped seed capsules. It grows on ledges of sea cliffs and islands along the coast of Labrador, northern and eastern Newfoundland, and the Maritimes. It is also found along northern coastal regions of Canada, Eurasia, Greenland, and Iceland.

Another species, Three-fingered Scurvygrass (*Cochlearia tridactylites* Banks), grows along the coast of the Gulf of St. Lawrence and is similar in appearance.

The young leaves and stems are picked just as flowering commences and can be used in a salad or sandwich, or steamed.

# Roseroot

*Rhodiola rosea* L.

*"Where abundant the plant is of great importance, owing to the scarcity of green vegetables in northern countries."*

(Fernald and Kinsey, 1943)

The fleshy rhizome of the Roseroot is said to smell like a rose.

The upright stems are 15-30 centimetres tall and have fleshy greyish leaves. There are separate male and female plants; both sexes have yellowish green flowers at their tips, but the male flowers have reddish tips. Roseroot grows on cliffs and ledges around the rim of the Arctic Ocean, and along the Arctic coasts of Canada, Alaska, and Eurasia. It occurs near the sea throughout Atlantic Canada, with the exception of Prince Edward Island.

Harvest the stems with leaves before the flower buds form. The young stems and leaves can be chopped into a salad or steamed and served with butter and pepper.

# Soapberry

*Shepherdia canadensis* (L.) Nutt.

*"Soapberry, though disagreeable to the white man's palate, furnished a very popular food to the Indians."*

(Fernald and Kinsey, 1943)

A spreading shrub which is usually less than 1 metre tall and has opposite leaves (1.5-5 centimetres long) that are oval with toothless margins. All parts of the shrub are covered with silvery tan scales.

Soapberry is found on headlands, barrens, and rocky slopes, and along the edges of woods and riverbanks. It is a boreal species found in New Brunswick, Nova Scotia, the northeast coast of Newfoundland, and northern Labrador. Soapberry is absent from Prince Edward Island.

The oval, reddish fruit are about 5-7 millimetres in diameter and ripen in mid-July. They contain saponin, which produces froth, like soap, when beaten with water. When sweetened with sugar and whisked, they produce a pink, cream-like suds, which is served as a dessert.

# 15

## CLEARINGS

*Much of the Atlantic region is covered by dense coniferous forest but there are open areas scattered through the woods. These clearings result from windfalls, forest fires, logging, and road building. Many edible plants grow in them because of their increased exposure to sunlight.*

# Squashberry

*Viburnum edule* (Michx.) Raf.

*"They had a pleasant acid flavour and tasted right well. Even if we had some other fruit, we should not have scorned these."*

(Kalm, 1748-1751)

This is a fairly short shrub, about 1-2 metres tall. The leaves are three-lobed and borne in pairs along the twig with fat winter buds at their bases. The fruit, which is 1-1.5 centimetres in diameter, is borne in clusters lower on the stem and has a single flat seed. It ripens late in August.

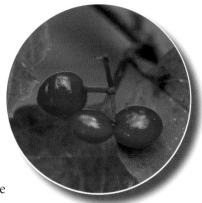

The squashberry is a circumboreal species that grows in open woodlands, particularly near their edges. It is found throughout Atlantic Canada, with the exception of Prince Edward Island.

The tart fruit is pleasant as a snack or made into a jelly.

# Highbush Cranberry

*Viburnum opulus* L. var. *americanum* Ait.

*"For richness of flavour, and for beauty of appearance, I admire the high-bush cranberries; these are little sought after, on account of the large flat seeds, which prevent them being used as jam: the jelly, however, is delightful, both in colour and flavour."*

(Traill, 1838)

This shrub is 1.5-3 metres tall. The leaves are in pairs and have three lobes, but the side lobes tend to be more pointed and narrower than those of the Squashberry. The clusters of red fruit are borne on the tips of the twigs and are about 1-1.5 centimetres in diameter. They remain on the bush into the winter.

The highbush cranberry grows in damp areas along ponds, roadsides, and woods. This species is found all across Canada, including the Maritimes and the island of Newfoundland, but it is absent from Labrador. It is not as abundant as the Squashberry.

The tart fruit, which ripen during the latter part of September, make a welcome snack while on an autumn stroll. It can also be made into jelly.

# Blueberry

*Vaccinium angustifolium* Ait.

*"Little hollow floures turning into small berries, greene at the first, afterward red, and at the last of a black colour, and full of a pleasant sweet [juice]."*

(Gerard, 1633)

Traditionally, in some parts of Newfoundland, the Blueberry has been called Hurts or Ground Hurts, depending upon its height. Hurts is an older spelling of Whorts, which is the name used for Blueberry in England.

This is a densely branched shrub (5-30 centimetres tall) that forms patches by underground stems. The leaves are narrow and oval-shaped, and, in autumn, they make hillsides seem ablaze when they turn scarlet.

The berries are 5-15 millimetres in diameter. Most berries have a whitish bloom which makes them appear pale blue but one variety has deep purple-black berries. They start to ripen in early August.

There are seven species in Atlantic Canada (species are listed in the Appendix; use Ryan [1978] to identify species). All have edible berries. They grow on heathlands and in open woods. This is a boreal species that is found throughout Canada. It occurs in the Atlantic region as far north as Nain in Labrador.

Blueberries can be prepared as a food in many ways: eaten fresh with cream; in pies, baked goods, jam, jelly, juice, and wine; and by the handful on a hike. They can be stored in the freezer or dried and used like currants in baking. Blueberries have been used traditionally in Europe to dye cloth.

# Skunk Currant

*Ribes glandulosum* Grauer

*"The fruit as big again as the ordinary red, but of a stinking and somewhat loathing savour: the leaves also are not without this stinking smell."*

(Gerard, 1633)

This sprawling shrub is about 0.5 metres tall with maple-like leaves. The leaves and fruit have a skunk-like odour. The bristly red berries are 5-10 millimetres in diameter and are borne in drooping clusters. They ripen in late July.

Skunk currants grow in damp woods and clearings of boreal regions throughout North America, including the Maritimes and Newfoundland and Labrador. This is the most common of the three currant species in the region, which also has a closely related native gooseberry (see Appendix; see Ryan [1978] for identification). All have edible fruit.

The "fragrance" of these berries is not evident in their jams and jellies. When making currant jam, add the sugar after cooking the fruit to avoid having tough "skins." For later use, store fruit in the freezer.

# Chuckley Pear

*Amelanchier bartramiana* (Tausch) Roemer

*"The Countrie fruites wild, are cherries small, whole groaves of them, a small pleasant fruites, called a Peare."*

(Mason, 1620)

This fruit has many common names such as Juneberry, Shadbush, Serviceberry, and Saskatoon. It sometimes goes by such names as Indian Pear, Pear Tree, and Wild Pear.

Chuckley Pears are shrubs or small trees ranging in height from 1 to 3 metres. The leaves are oval and have sharply toothed margins. The fruit (about 1.5 centimetres in diameter) are purple or reddish purple with a whitish bloom (some species have red fruit). They start to ripen in late July.

Chuckley Pears grow in clearings and woods and by streams and ponds throughout North America. This species occurs from Ontario to the Atlantic provinces, and north to Turnavik in Labrador. In the Atlantic region there are 11 species that bear a general resemblance to each other (see Appendix), with the exception of Newfoundland and Labrador, which is limited to six varieties. There are 15 species in North America. All have edible fruit.

Many of the fruit are infected by Cedar Apple Rust (*Gymnosporangium* spp.), a fungus which produces rusty orange finger-like fruiting bodies on the fruit, making them inedible.

The fruit are juicy and sweet with prominent seeds that soften when cooked. They can be stewed (3 cups fruit : 1 cup sugar) or made into pies, muffins, and other baked goods. They can also be dried and ground into flour for adding to cakes. Use a syrup pack (3 parts water : 1 part sugar) when freezing these fruit.

# Wild Strawberry

*Fragaria virginiana* Mill.

*"The fruit, not unlike to the Mulberrie, or rather the Raspis, red of color, having the taste of wine, the inner pulpe or substance whereof is moist and white."*

(Gerard, 1633)

There are two species of Strawberry in the Atlantic provinces. The Wild Strawberry occurs in boreal regions of North America, including the Maritimes, Newfoundland, and southern Labrador. The other, the Woodland Strawberry (*Fragaria vesca* L.), is a circumboreal species which occurs across the Maritimes, but is absent from Labrador and the Northern Peninsula of Newfoundland. Both species grow along the edges of woods, in clearings, and on grassy banks.

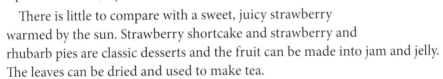

The plants, which are 10-20 centimetres tall, have a rosette of leaves, each with three leaflets. It spreads by stolons. The fruit are a rich, shiny red and have "seeds" on the surface in the Woodland Strawberry or sunken in pits in the Wild Strawberry. They ripen from mid-July onwards. Both are edible.

There is little to compare with a sweet, juicy strawberry warmed by the sun. Strawberry shortcake and strawberry and rhubarb pies are classic desserts and the fruit can be made into jam and jelly. The leaves can be dried and used to make tea.

# Pin Cherry

*Prunus pensylvanica* L.f.

*"The floures are small and white: after which come Cherries of a deepe red color when they be ripe, of taste somewhat sharpe, but not greatly unpleasant."*

(Gerard, 1633)

This is a small tree which grows up to 8 metres tall. It has narrow, pointed leaves which tend to hang downwards and turn scarlet in the autumn starting from the bases of the branches. The tree then looks like a glowing coal. The red fruit grow in clusters along the branches and are 4-8 millimetres in diameter. The fruit has a large hard-coated seed with a thin layer of flesh that starts to ripen in mid-August.

Pin Cherries grow in clearings, on burnt-over areas, and along shores and the edges of the forest. It is a boreal species of North America which occurs throughout Atlantic Canada north to southern Labrador.

The fruit are easily picked but large quantities are needed because of the thin flesh. They can be used to make jelly or wine (see Recipes).

# Chokecherry

*Prunus virginiana* L.

*"They so furre the mouth that the tongue will cleave to the roofe, and the throate wax horse with swallowing those Bullies."*

(Wood, 1634)

This is a small tree or large shrub about 2.5-5 metres tall. The leaves are broader than those of the Pin Cherry and do not hang downwards. The fruit are a dark purpleblack and are so astringent they cause the mouth to pucker.

Chokecherries grow on the banks of streams and along roadsides and the edges of forests. It is found across boreal North America, including the Maritime provinces. It is not as abundant in Newfoundland as the Pin Cherry, and is absent from the Northern Peninsula and Labrador.

The fruit are easy to pick and can be made into jam (strain out the seeds after cooking the fruit), jelly, and wine (see Recipes).

# Northeastern Rose

*Rosa nitida* Willd.

*"Euen children with great delight eat the berries thereof when they be ripe, make chaines and other pretty gewgawes of the fruit: cookes and gentlewomen make tarts and such like dishes for pleasure thereof."*

(Gerard, 1633)

There are six wild rose species in the Atlantic provinces, two of which occur in Newfoundland. The Northeastern Rose and the Virginian Rose (*Rosa virginiana* Mill.) are the most common species throughout Atlantic Canada. They can be distinguished by their stems – the Northeastern Rose has bristly stems like the canes of raspberry, and the Virginian Rose has only a few prickles on its stem like the Blackberry. Both species grow in peatlands, along roadsides, in wooded areas, and by riverbanks. They occur from Ontario to the Atlantic provinces but are absent from Labrador as well as the Northern Peninsula in Newfoundland.

The Northeastern Rose is a short shrub about 1 metre tall with stems densely covered with dark red bristles. The compound leaves have seven to nine leaflets. The fruit are bristly, red hips with "seeds" surrounded by a fleshy cup which ripen in late September.

The petals and hips of these two species and the cultivated varieties are edible.

Rose petals can be candied and used to decorate plates of baked goods and they can be made into vinegar to sprinkle on salads or fruit (2 cups petals, ½ cup brown sugar, 4 cups white wine vinegar; let stand for two weeks, strain, and bottle). The hips can be made into marmalade (see Recipes).

# Hairy Plumboy

*Rubus pubescens* Raf.

*"The fruit is among the most delicious productions of the uncultivated forest. To an agreeable combination of sweetness and acid it adds an aromatic fragrance."*

(Bigelow, 1817-1820)

Also called Ground Raspberry or Dewberry, the Hairy Plumboy is a short (15-20 centimetres tall) plant with a creeping stem that sends up fruiting stems. The leaves are often quite shiny. The fruit, which are 1-2 centimetres in diameter, resemble red Blackberries.

It is found throughout boreal North America in open, grassy areas and clearings in the woods.

The fruit has a rich tart taste and can be used to make pies, jam, and jelly.

# Canadian Blackberry

*Rubus canadensis* L.

*"The fruit or berry is like that of the Mulberry, first red, blacke when it is ripe, in taste betweene sweet and soure, very soft, and full of graines."*

(Gerard, 1633)

Blackberries often form impenetrable tangles of arching and trailing canes that are armed with formidable prickles. The compound leaves, which have five (three to seven) leaflets that radiate from the tip of the petiole, are also armed with prickles. The fruit, which are 2-3 centimetres in diameter, are black and ripen in September. The fruit in a cluster ripen over a period of time so repeated visits to a patch is needed to pick them all.

The Canadian Blackberry occurs west to Ontario. There are seven species (see Appendix; see Ryan [1978] for identification) reported in the Maritime provinces and three species in Newfoundland; it is absent from Labrador. They are similar in appearance and all have edible fruit. Blackberries grow in thickets and clearings and along roadsides.

The armed canes and leaves make picking painful but the fruit is a delicious prize. Enjoy them with sugar and cream or make jam, jelly, pies, and juice (crush fruit and strain). To make a cordial: use 2 cups juice and 2 cups sugar; bring just to a boil, let stand until cool, and decant (leave solids in pot). To serve, dilute to taste with mineral water, brandy, etc. The leaves can be dried and used to make tea.

# Raspberry

*Rubus idaeus* L.

*"Raspberries afford delicious distilled water [i.e., eau de vie], beer, mead and wine. Twigs dye silk and wool."*

(Rafinesque, 1830)

Raspberries form thickets of bristly canes about 1 metre tall. The compound leaves have three to seven leaflets. The canes, which last two years, produce fruit during their second growing season. The red thimble-shaped fruit are 1-2 centimetres in diameter.

This plant quickly produces patches in woods after a fire or logging. It also grows in fields, shrubbery, and roadsides. It is a circumboreal species that occurs throughout North America, including the Atlantic region north to southern Labrador.

The fruit are wonderful with cream or made into pies, jam, jellies, and wine. They freeze better than most fruit – freeze them in a single layer on a tray before packaging. Dried leaves can be used to make tea.

# Fireweed

*Chamerion angustifolium* (L.) Holub

*"The young leaves, under the name of* l'herbe fret, *are used by the Canadian voyagers as a pot herb."*

(Richardson, 1829-1905)

The reddish shoots of Fireweed are found among the remnants of last year's stalks. They are slender in overall appearance and have numerous narrow leaves on the stem.

Fireweed grows in disturbed habitats – cleared and burnt-over forests, fields, and roadsides. It grows in circumboreal regions of the Northern Hemisphere, including the Atlantic provinces.

Collect the shoots in late May or early June when they are about 15 centimetres high. Steam them and serve with butter, lemon, or drawn butter (see Recipes). The young flower stalks and leaves can be added to salads.

# Northern Wild Raisin

*Viburnum nudum* L. var. *cassinoides* (L.) Torr. & Gray

*"The cooked pulp has not proved attractive, but it is possible that by adding some tart fruit to it a palatable sauce might be prepared."*

(Fernald and Kinsey, 1943)

This shrub, which is 1-3 metres tall, is easily identified in any season. It has opposite, shiny leaves with few or no teeth along the margin. In winter, there are silvery brown winter buds, which are long and narrow. The clusters of starry, white flowers (spring) are followed by oval fruit, which change in colour from cream to pink to red (summer) and, finally, to purplish black (autumn).

Northern Wild Raisin grows in damp thickets, clearings, and forest edges, and by ponds and streams in eastern North America, from Ontario to Newfoundland and south to Tennessee. It is found across Newfoundland but only in a few sheltered locations on the Northern Peninsula, and it is absent from Labrador.

The fruit, which are 5-12 millimetres long, ripen in September, and when they have shrivelled they are sweet and make a nice nibble. The large seed makes them unsuitable for jam but juice can be made. The fruit taste like prunes and have a laxative effect.

# Dogberry

*Sorbus* spp.

*"Fruit very austere, never ripens ... makes a very strong cider."*

(Rafinesque, 1830)

This small tree (2.5-5 metres tall) has compound leaves with 11-17 leaflets. The orange fruit are 5-9 millimetres in diameter and ripen in September.

The circumboreal dogberry grows in thickets and open woods, and around clearings. There are four species (see Appendix) and one or more species occurs throughout Atlantic Canada. All have edible fruit.

The fruit are usually dry and bitter, but a variety exists in some areas, including the Burin Peninsula and South Coast of Newfoundland, that has a juicy, tart fruit like that of the Cranberry. The flavour improves after a frost. Wine and jelly can be made from the fruit and in Europe they are dried and ground into a flour which can be added to baked goods.

# Common Juniper

*Juniperus communis* L.

*"Linnaeus informs us ... that a fermented decoction of Juniper berries [is] used in Sweden as a common drink."*

(Bigelow, 1817-1820)

This sprawling evergreen, which is usually less than 1 metre tall, has prickly needles and forms thick mats or patches. The cones are berry-like (4-7 millimetres in diameter). When they are mature, which takes three years, they are blue-black with a white bloom.

This species is circumboreal and grows on the edges of woods and old fields and on heaths and hillsides throughout the Atlantic region. The other species, Trailing Juniper (*Juniperus horizontalis*) (see Appendix; see Ryan [1978] for identification), occurs in boreal North America and usually grows on headlands and hilltop heaths throughout the region. Both have edible berries.

The fruit provide the flavour for gin and can be used to flavour game, stuffings, marinades, and stews.

**Caution:** Do not eat while pregnant, as there are reports of it causing premature births.

# Mountain Alder

*Alnus viridus* (Chaix) DC subsp. *crispa* (Dryand.) Turrill

# Speckled Alder

*Alnus incana* (L.) Moench subsp. *rugosa* (DuRoi) Clausen

*"The young bark and winter-buds are popular nibbles with country boys, not alone for their tolerable flavor, but particularly for the beautiful, olive-brown saliva produced, which makes very emphatic spots on the lingering snow of early spring."*

(Fernald and Kinsey, 1943)

Alders are shrubs about 3 metres tall which have rough, wrinkled leaves. The two species can be distinguished by their leaves – Mountain Alder has seven to eight pairs of veins on each leaf and Speckled Alder has 10 or more pairs – and winter buds – Mountain Alder has pointed, oval buds attached by their base to the twigs and Speckled Alder has blunt, oval buds that are stalked (see Ryan [1978]).

Mountain Alder is found in boreal North America and northern Europe and grows in open woods and roadsides, and along streams and ponds. Speckled Alder is found in southern parts of the circumboreal region; it occurs in the Maritimes but is absent from the Avalon and Northern Peninsulas in Newfoundland, and it is restricted to the central part of southern Labrador.

Some people enjoy nibbling on the young bark and winter buds while on a winter hike.

**Mountain Alder (left and inset)**

# White Birch

*Betula papyrifera* Marsh.

*"The inner bark of various White Birches, ground to flour, has often been used as emergency bread-stuff."*

(Fernald and Kinsey, 1943)

This is our most common hardwood. It grows up to 20 metres and the bark ranges in colour from reddish brown in young trees to white in mature trees. It has broad, oval leaves and produces catkins with winged seeds in late summer.

White Birch grows in open woods and cut-over areas and along streams and ponds. It occurs throughout the Maritimes, Newfoundland, and in all but northern Labrador. It is a common species throughout the boreal regions of North America and a closely related species is found across the boreal regions of Europe and Asia. There are five dwarf birch species and two other birch tree species in the province (see Ryan [1978]).

Large trees can be tapped in the spring and the sap evaporated to make a syrup which has a more delicate flavour than maple syrup. The syrup can be made into wine or beer.

Yellow Birch (*Betula alleghaniensis* Britt., formerly *Betula lutea*; see Ryan [1978]) is found in a few sheltered sites along the southern part of Newfoundland, the Maritimes, west to Ontario, and south to Tennessee. Its young twigs are a source of oil of wintergreen.

# Beaked Hazelnut

*Corylus cornuta* Marsh.

*"When ground into meal they make a delicious cake-like bread comparable only to filbert bread."*

(Fernald and Kinsey, 1943)

This shrub can be easily mistaken for alder but it has distinctive beaked fruit rather than small cones. Each nut is surrounded by a fuzzy husk that has a long beak. The nuts, which are 1-1.5 centimetres in diameter, resemble the European Filbert and are found in clusters of two or three. The nuts are edible and ripen in early fall.

This species grows in clearings, along the edges of woods, and by streams and ponds across boreal North America, including the Maritimes; in Newfoundland it is absent from the Northern Peninsula as well as Labrador.

*55*

# FOREST FLOOR

*The boreal forest stretches in a broad band across northern North America, including Atlantic Canada. Several of our edible plants grow in the humus of the forest floor.*

**Left: Creeping Snowberry with Crackerberry and Clintonia.**

# Crackerberry/Bunchberry

*Cornus canadensis* L.

*"Some red [berries] which appear almost like coral ... which we ate delicious dishes when we could find them."*

(Sagard, 1624)

Arising from an underground stem, the plant, which is 5-15 centimetres tall, has a pair of tiny leaves lower on the stem and a cluster of four to six larger leaves at the top. The cluster of greenish flowers is emphasized by four white petal-like bracts. They are followed by orange berries (0.5-1 centimetre in diameter). The Swedish Bunchberry (*Cornus suecica* L.) is a closely related species. It can be distinguished by its flowers (dark purplish) and leaves (the lower leaves are similar in size to the upper leaves).

Crackerberry grows on the forest floor, in clearings, and on heaths. It occurs across boreal North America, including the Atlantic provinces north to Okak in Labrador, and eastern Asia. The Swedish Bunchberry is usually found on headland heaths in coastal Newfoundland, north to Nain in Labrador, and from Nova Scotia west to Alaska, as well as northern regions of Asia and Europe.

The berries make a refreshing snack for hikers or walkers. In Lapland, they are used to make a pudding: mix berries with whey, boil until thick, strain off seeds, and serve with cream.

# Creeping Snowberry

*Gaultheria hispidula* (L.) Muhl.

*"The leaves are used in infusion in place of tea and the berries are macerated in alcohol to make the 'liquer d'Anis.'"*

(Delamare et al., 1888)

There are many local names for this fruit, which is one of the greatest delicacies of the northern woods: Manna- or Magna-tea Berry, Maidenhair Berry, and Capillaire. The stems, which are slender and creep on the ground, have small (about 1 centimetre long) oval leaves along their length. The whole plant is covered with stiff rusty brown hairs. The white fruit (4-8 millimetres long) is a dry capsule full of seeds that is encapsulated by the calyx, which becomes fleshy. They ripen in August.

This species grows in boreal forests of North America, including Atlantic Canada north to Hopedale in Labrador.

The fruit and leaves have a wintergreen flavour. The fruit take considerable patience to pick as they hide beneath the leaves. Because of the numerous seeds, the fruit are best preserved whole (see Recipes). Use the freshly picked leaves to make tea as they lose their flavour when dried.

# Fiddleheads

Cinnamon Fern – *Osmunda cinnamomea* L.

Ostrich Fern – *Matteuccia struthiopteris* (L.) Todaro

Bracken – *Pteridium aquilinum* (L.) Kuhn

*"Because the spores cannot be seen by the naked eye, they were once believed to confer invisibility. Shakespeare alludes to this belief in* **Henry IV:** *'We have the receipt of fern-seed, we walk invisible.'"*

(Gordon et al., 1986)

Ferns grow throughout Atlantic Canada, mostly in moist habitats. The characteristic fiddleheads, which resemble the scroll of a violin, are the uncoiling young fronds found in the spring. Only a few species have fronds large enough to make them suitable for eating. Ostrich Fern, the classic fiddlehead, is common in New Brunswick, Nova Scotia, and Prince Edward Island, but much less abundant in Newfoundland and Labrador. The fiddleheads of Cinnamon Fern and Bracken can be collected in large quantities.

Pick stout young fronds when they are no more than about 15 centimetres high, break off the tender part, and strip off the woolly hairs or dry, papery scales. Wash them and prepare like asparagus: steam or boil until tender and serve with butter or hollandaise sauce.

**Ostrich Fern**

# Spruce

White Spruce – *Picea glauca* (Moench) Voss

Black Spruce – *Picea mariana* (Mill.) BSP

*"They make a kind of spruce beer of the top of the white fir which they drink in summer; but the use of it is not general and it is seldom drunk by people of quality."*

(Kalm, 1748-1751)

The two spruce and Balsam Fir are the dominant trees of our boreal forests in North America. They grow throughout the Atlantic region except for the northern tip of Labrador. The White Spruce grows on drier sites, while the Black Spruce grows mostly on wetter sites and bogs (see Ryan [1978] for identification).

The young shoots are gathered in the spring to make spruce beer. Resin oozes from the bark and forms crystalline clusters called myrrh, which has traditionally been used as chewing gum by youths and adults alike.

**White Spruce**

# Canada Yew

*Taxus canadensis* Marsh.

> *"The pulpy red portion of the berry is sweet and honey-like and perfectly edible."*
>
> (Fernald and Kinsey, 1943)

This short (less than 1.5 metres tall), evergreen shrub grows in small patches or as a single branch, which is easily mistaken for a branch of Balsam Fir (*Abies balsamea*). The dark green needles (up to 2.5 centimetres long) are flat and sharp-pointed. The fruit is a single dark green seed sitting in a bright red fleshy cup-like structure (about 1 centimetre wide).

It grows in woods, along streams, and at the edges of peatlands throughout eastern North America west to Manitoba, including Atlantic Canada north to central Labrador.

The fleshy cup of the fruit is juicy, slightly sweet, and edible but, beware, the seed is poisonous. Traditionally, branches of Canada Yew, called "palm" in Newfoundland, were used to decorate the dinner table on Palm Sunday.

# Corn Lily/Bluebead Lily

*Clintonia borealis* (Ait.) Raf.

*"The peculiar steel-blue berries are reputed to be poisonous, but so far as we can learn, no actual trials of them have been made."*

(Fernald and Kinsey, 1943)

The plant consists of a rosette of two to four large (10-20 centimetres long), oval leaves with a stalk rising from it (15-20 centimetres tall). At its apex the stalk bears two to eight pale greenish yellow flowers (1-2 centimetres wide) which are followed by metallic blue berries (about 1 centimetre in diameter).

It grows in woods from eastern North America west to Manitoba, including Atlantic Canada north to central Labrador.

The young leaves have a mild cucumber-like taste and can be added to salads. The leaves can also be harvested, just as they are unrolling, and used as a potherb. The berries are generally considered poisonous but this has not been established.

# Trailing Arbutus

*Epigaea repens* L.

*"The fragrant corollas of the Trailing Arbutus are spicy and slightly acid and are well known to children who eat them as a pleasant nibble and to relieve thirst."*

(Fernald and Kinsey, 1943)

Trailing Arbutus is a prostrate, evergreen shrub with oval leaves (2-6 centimetres long). The white flowers (about 1 centimetre wide) are tinged with rose and have hairs in their throats.

This species, which is the provincial floral emblem of Nova Scotia, grows in open woods east from Manitoba, including the Maritime provinces, western Newfoundland, and southern Labrador.

The flowers have a spicy, slightly acidic taste and can be nibbled on while walking.

# 71

# PEATLANDS

*Bogs and fens fill many of the depressions in our landscape.
This habitat is dominated by* Sphagnum, *a moss similar to
a sponge. A layer of* Sphagnum *is like a wet, springy carpet.
Patches of herbs, sedges, lichens, shrubs, and dwarfed trees
are scattered across the bog.*

# Cranberry

*Vaccinium macrocarpon* Ait.

*"The Marish Wortleberries ... little berries like unto the common blacke Wortleberries in shape ... in taste rough and astringent."*

(Gerard, 1633)

Cranberries have a threadlike stem that branches and creeps across the *Sphagnum* moss of the bog. The evergreen leaves are oblong with parallel sides and the margins may be slightly rolled under. The spherical berries, which ripen in late autumn, are 1-1.5 centimetres in diameter and are borne on a thin stalk that has two bracts above its middle. There is a leafy shoot beyond where the fruit is attached to the stem.

These plants tend to grow in seepage areas of coastal cliffs. They occur in bogs throughout the Maritimes and other areas of eastern North America. Cranberries grow in Newfoundland but not in Labrador.

Cranberries were probably the first native North American fruit eaten in Europe as records show that tons were shipped there in barrels of water. The fruit makes a tart sauce to accompany holiday fowl, relish (see Recipes), jelly, beverages, pies, and other baked goods. Cranberry syrup can be whipped with gelatine or egg white to make a light dessert.

# Marshberry

*Vaccinium oxycoccus* L.

*"Toca [native name] were put into little cakes or eaten raw."*
(Sagard, 1624)

The Marshberry is an evergreen with threadlike creeping stems. The leaves (about 6-7 millimetres long) are oval and broader at the base, with the margins rolled under. The oval berries (1-1.5 centimetres in diameter) are borne at the tips of the stems. The two bracts of the fruit stalk are at or below the middle of the fruit stalk. The fruit ripen in late fall but the flavour improves after several frosts.

This circumboreal species is absent from the Maritimes but grows on peatlands across Newfoundland north to Hopedale in Labrador.

The fruit is used to make jam, jelly, wine, and baked goods.

# Bakeapple

*Rubus chamaemorus* L.

*"The fruit quencheth thirst, being eaten as Worts [blueberries] are, or the decoction made and drunke."*

(Gerard, 1633)

Bakeapples grow about 15 centimetres tall and have one to three five-lobed leaves, which are a rough dark green. There are separate male and female plants. They flower in June, and late frosts often damage the flowers and cause crop failures. More abundant crops are found near the coast and on islands where they escape late frosts. Each plant has a single berry (1.5-2 centimetres in diameter) which ripens in mid-August.

This is a circumboreal plant of peatlands. It is found throughout the Maritimes and Newfoundland and Labrador.

The fruit has superior flavour when fully ripe – a translucent honey-gold colour. Harvesting bakeapples is a back-breaking task but one that brings ample reward. The soft nature of the fruit results in much juice in the bucket. They are usually served as a jam in a dish or pastry shell with cream or as a sauce on cheesecake. Wine and liqueur are also made from bakeapples.

# Labrador Tea

*Rhododendron groenlandicum* (Oeder) K. & J.

*"The English in these parts [Hudson Bay] make a Drink of itt ... being of a fine flavour, and Reckon'd Very wholesome."*

(Isham, 1743)

This shrub grows up to 1 metre in height and has distinctive evergreen leaves – a wrinkled dark green upper surface with the margins rolled under to the lower surface, which is covered with a mat of rusty or golden brown hairs.

It grows throughout boreal North America on peatlands and in damp woods.

The leaves are used to make tea and are collected in early spring. So as not to weaken the plants, pick a few from each plant and store them in the freezer or dried in an airtight container. Other related species can also be used to make tea: the leaves and flowering tips of Lapland Rosebay (*Rhododendron lapponicum* (L.) Wahlenb.); the Ojibwa used leaves of Bog Rosemary (*Andromeda glaucophylla* Link) and Leatherleaf (*Chamaedaphne calyculata* (L.) Moench).

**Caution:** To make tea, do not boil the leaves as this releases the harmful alkaloid andromedotoxin. After boiling the water, remove it from the heat and steep leaves for 5-10 minutes and serve with honey and/or lemon.

# Sweet Gale

*Myrica gale* L.

*"They [in Quebec] put the leaves into their broth to give it a pleasant taste."*

(Kalm, 1748-1751)

Sweet Gale is a short shrub (about 1 metre tall) that has glands which exude an aromatic resin on the leaves, young twigs, and fruit. Hold a twig up to the light and look for tiny honey-like globules. There are separate male and female plants. The fruit is a cluster of angular green nutlets.

This is a shrub of damp habitats – peatlands and by ponds and streams. It is a boreal species of North America and Europe that grows throughout Atlantic Canada north to Nain in Labrador.

Collect the leaves during the growing season and the fruit in the autumn. The leaves can be used like bay leaves in soups and stews or for tea. The fruit has a peppery, sage-like taste and is used to season roasts or nibbled on while on a hike.

**Left: Sweet Gale (male) Inset: Sweet Gale (female)**

# Blue Marsh Violet

*Viola cucullata* Ait.

*"Many herbes in the spring time there are commonly dispersed throughout the woods, good for broths and sallets, as Violets, Purslin, Sorrell, etc."*

(Smith, 1612)

This violet grows to about 15 centimetres tall. It has a rosette of heart-shaped leaves from which the blue flowers (1.5-2 centimetres broad) arise.

This species is the provincial floral emblem of New Brunswick. It grows in eastern North America from Ontario south to Georgia. There are 15 species of violets found in areas of the Maritime provinces (with the fewest varieties occurring in Prince Edward Island), and 12 species in Newfoundland and Labrador. Most have blue flowers but a few have white or yellow ones. Some grow as rosettes of leaves or with leafy stalks. They all have similarly shaped leaves and typical violet flowers, and all are edible.

Young leaves are high in vitamin C and can be added to salads as can the petals of the flowers. The flowers, when candied (see Recipes), are used to decorate cakes or plates of baked goods.

# Northern Fly Honeysuckle

*Lonicera villosa* (Michx.) R. & S.

*"Although but little known as edible fruit, the Waterberries, as they have been appropriately named in eastern Maine, are delicious, in flavor somewhat suggesting blueberries."*

(Fernald and Kinsey, 1943)

This is a short shrub (up to 1 metre tall) that has oval leaves with toothless margins and a bluish appearance. The pale yellow flowers, which have a lemon fragrance, are followed by dark blue, edible berries.

It grows in damp habitats – peatlands, wet meadows, and by ponds and streams. It is a species of boreal North America which occurs throughout Atlantic Canada north to Goose Bay in Labrador.

The fruit, which ripen in late July, are rarely abundant, so enjoy them as a nibble while on a hike.

# Huckleberry

*Gaylussacia baccata* (Wangenh.) Koch – Black Huckleberry

*Gaylussacia dumosa* (Andr.) T. & G. – Dwarf Huckleberry

*"An important food, dried the berries for winter use, ate fresh, made into drinks."*

(Bye, 1970)

Huckleberries closely resemble the blueberry in its leaf, flower, and fruit but can be distinguished by the globules of resin which dot the whole plant. The globules can be seen when the plant is held up to the light.

The two species grow in eastern North America on bogs and in damp woods throughout Atlantic Canada, but are absent from Labrador and the Northern Peninsula of Newfoundland.

The fruit are black instead of the frosted blue of blueberries and ripen in late summer. They are edible but have large seeds, so use them to make jelly or juice.

**Dwarf Huckleberry (left and inset)**

# Chokeberry

*Photinia floribunda* (Lindl.) Robertson & Phipps – Purple Chokeberry

*Photinia melanocarpa* (Michx.) Robertson & Phipps – Black Chokeberry

*"The raw berries have a good flavor but are very puckery, much as choke-cherries. It is stated, however, that the Indians used these fruits, destroying the puckery quality by cooking."*

(Fernald and Kinsey, 1943)

This is a shrub of peatlands and damp areas. It is about 1 metre tall, with leathery, shiny leaves, which are oval and broadest towards the tip, and black or red fruit. The whole plant has some degree of fuzzy hairs present.

Chokeberries grow throughout eastern North America. The two species (see Ryan [1978] under their former name, *Aronia*) are found across Atlantic Canada, but are absent from Labrador and the Northern Peninsula of Newfoundland.

The fruit ripen in August and are edible but not particularly palatable, as the common name suggests.

**Black Chokeberry (left and inset)**

# 91

## SEASIDE

*A band of plants which thrive on beaches
just above high tide are "strand plants,"
which have their seeds dispersed by ocean currents.
Our strand plants are found on the shores
of the north Atlantic Ocean.*

**Seaside Plantain**

# Sea Rocket

*Cakile edentula* (Bigel.) Hook.

**"With a flavor reminiscent of mustard green and horseradish with a hint of garlic."**

(Gibbons, 1964)

Sea Rocket forms a sprawling patch about 0.5 metres in diameter. The bright green leaves (3-6 centimetres long) are thick, broader towards the tip, and have rounded teeth along the edges. The four-petalled white flowers, which are produced throughout the summer, are succeeded by a fruit that resembles an ice cream cone.

This is a strand plant. It grows on sandy and rocky beaches above high tide in northern Europe, and on the east coast of Canada north to Forteau in southern Labrador.

The stems, leaves, seed pods, and flower heads are edible throughout the growing season. The whole plant can be steamed and served with butter. The leaves and capsules can be used in salads and the leaves in sandwiches.

# Seaside Plantain

*Plantago maritima* L.

*"Seaside Plantain is not very generally known as one of the most available summer vegetables, but on the New England coast, especially by the fishermen of eastern Maine, and in Nova Scotia, where the plant is regularly gathered under the name of Goose-Tongue, it is extensively used."*

(Fernald and Kinsey, 1943)

Seaside Plantain, also called Goosetongue, has long, narrow, fleshy leaves (10-15 centimetres long) in a rosette. The flowering stalks arise from the centre and the shaft of flowers and fruit looks like a rat's tail.

This species, which occurs along the coasts of North America, South America, and Eurasia, can be found in the cracks of cliffs, on gravel slopes, and on beaches.

Collect the leaves, a few from each plant. Chop the younger leaves and add to salads; the larger leaves can be treated like string beans: cut, steam for about 15 minutes, and serve with butter.

# Scotch Lovage

*Ligusticum scothicum* L.

*"In America Scotch Lovage has never come into general favor, but in the Hebrides and other maritime sections of Scotland it has long been used, either as a cooked potherb or, when blanched by covering with litter [leaves and twigs that allow lovage to emerge from the soil], the leaf-stalks have found use as a substitute for celery."*

(Fernald and Kinsey, 1943)

This species is sometimes called Alexanders. The plant ranges in height from 30 to 60 centimetres. The compound leaves are stiff and shiny. The flat heads of small (2-3 millimetres in diameter) white flowers (umbels), produced in mid-summer, are followed by ribbed fruit (about 5 millimetres long).

Scotch Lovage grows on cliffs and beaches along the coast of Atlantic Canada and the northern shores of Quebec and Ontario (Hudson's Bay). It is also found on the coasts of northern Asia and Europe, including the Hebrides and Scotland.

This plant tastes like celery and parsley. Chop the young petioles and leaves and add to salads, or steam them as a potherb and serve with butter. The young shoots can be candied like Angelica (see Recipes).

# Beach Pea

*Lathyrus japonicus* Willd.

*"Dr. Harold St. John, returning from a summer on the Labrador coast, stated that there the pips or young shoots coming through the sand and looking like the shoots of garden peas, (but often red or purple), are gathered and boiled and after cooking are made into salads."*

(Fernald and Kinsey, 1943)

This member of the Pea Family forms circular mats of leaves, flowers, and fruit. The compound leaves have three to six pairs of broad leaflets and tendrils at their tips. The purple-pink flowers are about 2 centimetres long and are followed by green, pea-like pods (3-7 centimetres long).

Beach Pea grows on rocky beaches of coastal Alaska, Canada, and Eurasia. It is found throughout the Atlantic provinces north to Nain.

Collect the pods while the seeds are young and tender, shell, boil the peas for 15 minutes, and serve with cream and black pepper.

# Strand Wheat

*Ammophila brevilingulata* Fern. – American Beachgrass

*Leymus arenarius* (L.) Hochst. – Strand Wheat

**"Ever since the eleventh century the Strand-Wheat has formed a staple cereal of the Islanders."**

(Fernald and Kinsey, 1943)

These are sturdy grasses found on the dunes of sandy beaches. They grow 0.5-1 metre tall and produce heads (up to 15 centimetres long) of large grain (about 1 centimetre long).

Strand Wheat, a native of Europe, has been introduced into southern British Columbia, eastern Canada, the Maritimes, and northeastern Newfoundland. American Beachgrass is found on the shores of the Great Lakes and the coasts of eastern Canada, but is absent from Labrador and the Northern Peninsula of Newfoundland.

The mature grain is collected and ground into flour. Try the flour in a batch of scones and enjoy the flavour of wild grain.

**Strand Wheat**

# Orach

*Atriplex glabriuscula* Edmondston – Smooth Orach

*Atriplex patula* L. – Orach

*"As a potherb Atriplex is superior to Lamb's-Quarters or Pigweed. The succulent leaves or young tips, especially when the plant grows along the seashore, are juicier and somewhat impregnated with salt."*

(Fernald and Kinsey, 1943)

The Orachs resemble some of the Knotweeds and Lamb's Quarter in growth form – their wiry stems are branched and have arrowhead-shaped leaves and clusters of knobby reddish green fruit.

Both species grow on coastal beaches around northern North America and northern Europe. Smooth Orach reaches as far north as Red Bay in southern Labrador.

The leaves and young stems are used as a potherb.

**Atriplex patula**

# 105

## BANKS AND SHORES

*Some plants grow in damp soil or in shallow water
on the banks of rivers and the shores of ponds.
Since water may be contaminated with microorganisms,
care should be taken in collecting these plants for
consumption. Pick those parts of the plant
above the water's surface and
wash them well before eating them.*

# Mint

*Mentha arvensis* L. – Wild Mint

*Mentha cardiaca* Baker – Heart Mint

*"The whole herbe is of a pleasant smell. Mint is marvellous wholesome for the stomacke."*

(Gerard, 1633)

Both mints have opposite leaves and stems which are square in cross-section. These are characteristics of the Mint Family. Wild Mint grows up to 60 centimetres tall and the whole plant is covered with short, stiff hairs. The lilac flowers form a corona at the bases of each pair of leaves. Heart Mint has hairless, broad, dark green leaves and purple stems. Both species are edible.

Mints grow in wet soil and along streams and ponds. Wild Mint is found across boreal North America, including the Maritimes, Newfoundland, and Labrador north to Goose Bay. Heart Mint was introduced from Europe and grows in British Columbia and Ontario east to the Maritimes, as well as western and eastern Newfoundland.

The plants can be harvested throughout the summer. Cut the stems above the waterline to avoid contamination. Fresh leaves can be dried to make tea; they can be added to salads and are a feature of several Middle Eastern dishes; and they can be used to make sauces and jellies (see Recipes). Mint helps in the digestion of lamb and veal. Mint confections are a traditional ending for a meal.

**Wild Mint**

# Watercress

*Nasturtium microphyllum* Boenn.

*"Early settlers brought the plant to America chiefly because of its effectiveness in preventing scurvy, for the plant is rich in vitamin C."*

(Gordon et al., 1986)

Watercress grows along the edges of streams and ponds. The stems sprawl on mud or float in water. The leaves have three to nine oval lobes with a larger terminal lobe. The white flowers are about 5 millimetres in diameter and the slender seed pods are 10-25 millimetres long.

Watercress was introduced from Europe and grows in boreal regions of Canada, including the Maritimes and eastern Newfoundland.

Pick the tips of the stem above the water to avoid contamination or grow this plant in a damp spot in the garden. The leaves and seed pods have a sharp, peppery taste. Use them in salads and sandwiches, as a potherb, in soup, or stir-fried with grated, fresh gingerroot.

# Waterlily

*Nymphaea odorata* Dryand – Fragrant Waterlily

*Nuphar variegata* Durand – Bullhead Lily

*"The fresh leaves [of Nymphaea] ... in Canada they are eaten in the spring, boiled for greens."*

(Rafinesque, 1830)

The two common species of waterlily can be distinguished by their leaves – Bullhead Lilies have oval, olive green leaves (10-25 centimetres wide) and Fragrant Waterlilies have more circular leaves (10-20 centimetres wide) that are dark green with dark red undersurfaces. The flowers of the Bullhead Lily are yellow globes (2.5-4 centimetres in diameter) and those of the Fragrant Waterlily are double white blossoms (about 19 centimetres in diameter).

These aquatic plants have thick rhizomes that are rooted in the mud of pond bottoms and leaves that float on the surface. The Fragrant Waterlily grows in eastern North America; it is found throughout Atlantic Canada, but is absent from Labrador and the Northern Peninsula of Newfoundland. The Bullhead Lily is found across the boreal regions of Canada and in the Atlantic provinces as far north as central Labrador.

The Ojibwa cooked the flower buds of *Nymphaea* and other First Nations people boiled the rhizomes of *Nuphar.*

**Fragrant Waterlily**

# Cattails

*Typha latifolia* L.

*"In different regions the Cat-tail has won considerable attention as a food-plant, although it is noteworthy that its reputation varies in different places."*

(Fernald and Kinsey, 1943)

Cattails have strap-like leaves (1-2 metres long) and flowers crowded into a dark brown sausage shape (female flowers) with a plumed spike above (male flowers).

This species grows from Alaska to eastern North America, and is also found in Eurasia and North Africa. It was introduced to the island of Newfoundland from adjacent parts of Canada and is spreading eastward in the ditches of the Trans-Canada Highway. There are also populations in swampy areas of communities along the Highway. It has not been found in Labrador.

The rhizomes are used as a source of starch and the immature flower spikes are eaten like corn on the cob. Collect the spikes when they are green, that is, when they are about to break through the papery sheath that encloses them. Remove the sheath, boil for a few minutes in salted water, and serve with butter. The core is inedible.

# Marsh Marigold

*Caltha palustris* L.

*"The Marsh-Marigold ... has long been one of the most popular spring greens of New England. As early as 1784, the Massachusetts botanist, Manasseh Cutler, spoke of it as an esteemed potherb."*

(Fernald and Kinsey, 1943)

Marsh Marigolds are 20-40 centimetres tall and have broad (6-12 centimetres) kidney-shaped leaves. The yellow flowers (1.5-4 centimetres wide) resemble starry buttercups.

They grow in damp meadows and wet depressions in woods, and along streams and ponds in boreal Eurasia and North America. In Atlantic Canada, they occur in the Maritime provinces, western Newfoundland, and southern Labrador.

The flower buds can be boiled, pickled in vinegar, and used as mock capers. Traditionally, the petals were used to colour butter. The leaf blades (remove petioles) are used as a potherb – boil in two waters to get rid of the bitter taste.

**Caution:** Do not eat the raw leaves of Marsh Marigold as they contain the poison helleborin, which can be removed by boiling.

# Cleavers

*Galium* spp.

*"Women do vsually make pottage of Cleuers with a little mutton and oatemeale, to cause lanknesse, and keepe them from fatnes."*

(Gerard, 1633)

There are 11 species of Cleavers in Atlantic Canada. Some species are upright but most have weak stems and sprawl along the ground. The stems are 10-30 centimetres long and square in cross-section, and have short spines which curve towards the base of the stem and whorls of four to eight leaves. The white flowers are tiny – most are 2-4 millimetres in diameter.

Cleavers usually grow in moist meadows and other damp habitats in boreal regions of the Northern Hemisphere.

Young shoots are used as potherbs, and it has been suggested that the seeds, when dried and roasted, substitute for coffee.

# 119

## DISTURBED AREAS

*Around communities, these are sites which have been cleared of the original vegetation. "Weeds" grow throughout – in gardens, lawns, roadsides, and parking lots, etc. Weeds are plants that have been introduced from other places by our activities – they can travel in and on vehicles, cargo, hay and other animal feed, etc. Most of our weeds are from Europe, where many have provided food for centuries.*

**Curled Dock**

# Japanese Knotweed

*Fallopia japonica* (Houtt.) Ronse-Decr.

*"Steamed or boiled for four minutes they become as soft as cooked rhubarb and are delicious, especially when chilled and dressed as a salad."*

(Fernald and Kinsey, 1943)

A perennial herb, Japanese Knotweed is a plant of a thousand names – Mexican Bamboo, Mile-a-Minute, September Mist, etc.

The shoots are olive green with considerable red pigment. The bamboo-like stems grow each year to a height of 1-3 metres and have oval, olive green leaves.

This native of Japan and the adjacent Asian continent was introduced and has spread throughout many regions of North America, including the Maritimes and Newfoundland. A closely related species, Giant Knotweed (*Fallopia sachalinensis* (Schmidt) Ronse-Decr.), which resembles Japanese Knotweed, is found in British Columbia, Ontario, the Maritimes, and southern and western Newfoundland. Both varieties are edible.

The young shoots rise among the stalks of the previous year. Collect them when they are about 15 centimetres tall and the leaves are starting to unroll. Prepare like asparagus: snap off the tender tip, boil or steam, and serve with butter and salt. Eat in moderation, initially, as they may be slightly laxative.

# Stinging Nettle

*Urtica dioica* L.

*"Spring shoots are boiled in Europe for pot herbs."*

(Rafinesque, 1830)

The stems grow up to 1.5 metres tall and have lance-shaped, toothed leaves in pairs. The whole plant is covered with stinging hairs.

Stinging Nettles thrive on rich organic sites such as around barns and homesteads and around wharves. Stinging Nettle is located throughout North America, and a similar species, Burning Nettle (*Urtica urens* L.), is found across Canada. Both types are edible.

The young leaves can be blanched and used as a spinach substitute, and the young shoots as a potherb or in soup. The leaves can be dried to make tea. The stems are quite fibrous and, with a felt of stiff hairs, it is a chewy herb. The stems were processed like flax by native North Americans and Scots to make rope and cloth.

**Caution:** Harvesting is best done with stout leather gloves and scissors to avoid being stung. The burning can be neutralized with the sap of Patience Plant (*Impatiens*) or Dock (*Rumex crispus*).

# Sorrel

*Rumex acetosa* L. – Garden Sorrel

*Rumex acetosella* L. – Sheep Sorrel

*"The juice hereof [Garden Sorrel] in Sommer time is a profitable sauce in many meats, and pleasant to the taste."*

(Gerard, 1633)

Sheep Sorrel grows about 30 centimetres tall and has arrow-shaped leaves and clusters of spherical red-tinged flowers (1-2 millimetres in diameter). Garden Sorrel grows about 50 centimetres tall and has similar appearing flowers and lance-shaped leaves.

The sorrels are introduced species. Sheep Sorrel grows in gardens and disturbed sites throughout North America, including Atlantic Canada north to central Labrador. Garden Sorrel has a similar distribution, except in Labrador, where it is found in southern areas.

Sheep Sorrel leaves are a favourite with children, who know them as Sweet Leaf, Laddie Suckers, and Sally Suckers. The leaves of both sorrels are used in salads, sandwiches, and soups. They can be chopped and used as a garnish on fish. In the Middle Ages, Garden Sorrel leaves were mashed, mixed with vinegar and sugar, and used as a sauce on cold meat. Note: the sharp taste is due to oxalic acid, so sorrels should be avoided by those with gout, rheumatism, and kidney ailments.

**Sheep Sorrel**

# Curled Dock

*Rumex crispus* L.

*"Farmers ... generally boil the leaves in the water in which they had cooked meat. Then they eat it alone or with meat."*

(Kalm, 1748-1751)

Curled Dock grows about 80-100 centimetres tall, has narrow oval leaves (up to 30 centimetres long), and clusters of fruit which look like dry tea leaves. Children use the fruit as "tea" when they play house.

This species was introduced from Europe. It grows in disturbed areas throughout boreal regions of North America, including the Maritimes, Newfoundland, and southeastern Labrador.

Collect the leaves in spring and use as a potherb. If the flavour is too strong, boil in two waters.

# Red Clover

*Trifolium pratense* L.

*"The leaves ... hauing for the most part in the midst a white spot like a halfe moon."*

(Gerard, 1633)

Red Clover grows in clumps about 30-40 centimetres tall. The leaves have three leaflets and each leaflet has a white chevron mark. The claret-red flowers are in rounded heads about 2 centimetres in diameter. There are two other edible species of clover: Alsike Clover (*Trifolium hybridum* L.) has a similar growth habit to Red Clover, but the head has pink and white flowers, each of which has a thin stalk, while those of Red Clover do not. White Clover (*Trifolium repens* L.) has a creeping stem and a flattened head of white flowers.

The three clover species grow in lawns, gardens, and disturbed areas throughout North America, including Atlantic Canada north to central Labrador.

Young parts of the plants can be added to a salad or used as a potherb. Flowers can be used as a sandwich filling or added to a salad. Fresh flowers can be used to make tea or dried at room temperature to preserve their flavour and stored in a jar. The flowers can be used alone or mixed: 1 part dried mint leaves to 4 parts dried clover blossoms. Steep and serve with honey.

# Dandelion

*Taraxacum officinale* Weber

*"What Virtues this common Herb hath, that's the reason the French and Dutch so often eat them in the Spring."*

(Fernald and Kinsey, 1943)

Dandelions have a long taproot, long narrow leaves with jagged teeth along both edges, and disc-like deep yellow flowers.

This plant grows wherever we are. It is found around communities throughout many parts of the world, including Atlantic Canada north to central Labrador.

It could be argued that Dandelion and Coltsfoot are responsible for saving many of our ancestors' lives. In earlier times, when people became ill in late winter from vitamin and mineral deficiencies, physicians dug and administered Dandelion. Coltsfoot (*Tussilago farfara* L.) took care of respiratory complaints.

Leaves, collected in early spring, are added to salads. Collect plants before or just as they begin to flower, wash thoroughly, boil leaves and flower buds for 5 minutes, and serve with butter and salt or crisp bacon. Petals can be added to omelettes, and flowers can be made into wine (see Recipes). The roots make a coffee substitute: dig roots in spring, wash, roast slowly in oven (about 180°F) until they are dark brown and snap (about 4 hours), grind, and prepare as for coffee. The puffballs of seeds are blown to predict the number of events in one's life.

# Lamb's Quarter

*Chenopodium album* L.

*"The* **Chenopodium album** *L. which grew in great quantities in rich soil and was the second plant used as kale. Only the young plants, a few inches in height were used."*

(Kalm, 1748-1751)

At maturity, this plant can reach 0.5 metres or more in height and the leaves are shaped like an elongated diamond. The green fruit are triangular in shape. The leaves of young plants appear to be dusted with sugar.

Lamb's Quarter grows in gardens and disturbed areas throughout Eurasia and North America, including the Maritimes and Newfoundland, but is absent from Labrador.

This plant should be picked when young (about 15 centimetres tall). The whole plant makes an excellent potherb and can be cooked and served like spinach.

# Chickweed

*Stellaria media* (L.) Villars

*"Little birds in cadges (especially Linnets) are refreshed with
the lesser Chickweed when they loath their meat [seeds]."*

(Gerard, 1633)

The trailing branches radiate from a slender taproot and produce a circular mat up to 0.5 metres in diameter. The lower leaves have petioles, while the upper leaves do not. Each of the five white petals in the flower is deeply lobed, giving the appearance of ten petals.

Chickweed, which was introduced from Europe, grows in disturbed ground across North America. It occurs in the Maritimes, Newfoundland, and north to Hebron in Labrador.

Chickweed growing as weeds in one's garden are often the lushest as they flourish in the rich soil. Harvest young, vigorously growing tips and add to a salad or steam and serve with butter. Stems can be mixed with other greens, especially strong-flavoured ones, to mellow the taste.

# Shepherd's Purse

*Capsella bursa-pastoris* (L.) Medic.

*"A good substitute for spinach. Delicious when blanched and served as a salad. Tastes somewhat like cabbage but is much more delicate."*

(Medsger, 1939)

The rosette of hairy leaves is about 15 centimetres in diameter. The flowering stalk has tiny white flowers which are followed by the distinctively shaped seed capsules from which the common name arises.

Introduced from Europe, this species grows in yards and disturbed sites of communities across North America, including Atlantic Canada north to Nain in Labrador.

The plants are collected when they are young, before or just as they start to flower. The peppery young leaves can be eaten like spinach, added to salads, or used to season stews. The seeds can be substituted for mustard.

# Winter Cress

*Barbarea vulgaris* Ait.

*"In winter when salad herbes bee scarce, this herbe is thought to be equall with Cresses of the garden, or Rocket."*

(Gerard, 1633)

This is a winter annual – a rosette of leaves (15-20 centimetres in diameter) which grows in the autumn, overwinters, and flowers in the spring. Most annual plants overwinter as dormant seeds in the soil. The leaf segments are rounded and the petiole has several lobes of leaf blade. The yellow flowers are followed by rod-shaped seed capsules which release their seeds in late summer and restart the cycle.

Introduced from Europe, Winter Cress grows on roadsides and other disturbed sites throughout North America, including Atlantic Canada north to central Labrador.

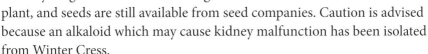

The genus name, *Barbarea*, was given since it was "the only green plant that could be gathered and eaten on Saint Barbara's Day, which falls on the fourth day of December" (Gibbons, 1962). There is a very long tradition of consuming this plant, and seeds are still available from seed companies. Caution is advised because an alkaloid which may cause kidney malfunction has been isolated from Winter Cress.

The centre leaves of the rosette are collected before it flowers and used in salads or as a potherb. The flowering buds, which taste like a superb broccoli, are steamed and served with butter and salt.

# Pennycress

*Thlaspi arvense* L.

*"The young leaves are edible, tasting somewhat mustard-like, with a suggestion of onion."*

(Fernald and Kinsey, 1943)

Pennycress is an annual which has two growth forms. It can grow 10-40 centimetres tall or, often, its stems radiate from the root to form a mat on the ground. The leaves (2.5-6 centimetres long) have a smooth margin or a few teeth and clasp the stem at their bases with two "ears." The white flowers (about 3 millimetres in diameter) are followed by flat, circular fruit (10-15 millimetres long).

This species was introduced from Europe and grows on roadsides, parking lots, and other disturbed sites across North America, including Atlantic Canada north to central Labrador.

The young leaves, which have a sharp, mustard-like flavour, are chopped and added to salads or steamed as a potherb.

# RECIPES

*This section contains a few basic recipes
for which edible fruits and herbs can be used.*

General indications for serving each plant are given in its description. Many of the recipes are those of my grandmother, who was from Bonavista North in Newfoundland, or those that I have adapted from general instructions in various references. Several have been kindly passed on by friends.

You do not need lots of berries for most recipes – a half cup is enough for a sauce, cottage pudding, or muffins, or several types of herbs can be steamed together for a meal of greens. You only need a few leaves or flowers to make a cup or pot of tea.

The following recipes apply generally to herbs and fruits. Any differences or special recipes are mentioned in the descriptions of each species.

# Potherbs

Plants that can be used as a potherb are picked at an early stage in the spring, usually before they bloom, as they become bitter when mature. After cleaning them of soil and damaged leaves, etc., cook them in a steamer. This preserves more of the flavour and texture than boiling. Serve with butter or drawn butter. Some potherbs, like dandelions, have a strong taste and may be served with chopped bacon or olive oil heated with chopped garlic and fresh gingerroot.

# Drawn Butter

This sauce is traditionally used on fish but can also be used on potherbs.

> 1 onion, chopped
>
> ⅓ cup butter
>
> 2 cups water
>
> ¼ tsp salt
>
> 2 tbsp all-purpose flour

Cook onion in butter until soft. Add water and salt and bring to a boil. Mix the flour with ¼ cup water and add to onion mixture. Boil, while stirring, until the sauce has thickened.

# Tea

Leaves, flowers, or fruit can be used to make tea. These can be used fresh in some cases but, often, the plant material is dried. The substance which gives the flavour to the tea is often a volatile oil, so the method of drying and storing is important. Collect the part of each species to be used and dry at room temperature in a well-ventilated room. Store in a jar or tea tin.

The traditional method for making tea starts with heating the teapot by swirling with hot water. Discard water, place the ingredients for the tea into the pot, and then pour in boiling water just removed from the heat. Cover the pot with a tea cozy and let it sit for 5-10 minutes. Serve plain or with lemon, honey, or milk, etc. The important point is not to boil the tea as this can cause some of the undesirable substances to dissolve.

# Candied Plants

Several plants may be candied and used in baking, like fruit peels at Christmas.

# Scotch Lovage

Cut the young leaf stalks into short segments and boil in water for 15-20 minutes. Make a syrup by boiling 1 cup of sugar and ¾ cup of water. Strain the stalks and drop into the syrup. Boil for 10 minutes. Drain on waxed paper, roll in sugar, and dry for 2 days. Use as candy or in baking.

# Whole Fruit

The Creeping Snowberry is best preserved whole in a syrup because of all the seeds it contains.

> 1 cup fruit
>
> 1 cup sugar
>
> 1 cup water
>
> 2 slices lemon

Boil the sugar, water, and lemon for 5 minutes. Add the fruit and simmer until the fruit appear clear. Seal in hot sterilized jars.

# Wine

Wine-making is a rewarding art. There are more technical methods of wine-making but I have included two general recipes – one for flowers and two for fruit.

## Dandelion Wine

This basic recipe can be modified to your taste. Pick a gallon of flowers on a sunny day, put in a crock, and pour 1 gallon of boiling water over them. Cover and leave at room temperature for 3 days. Strain through a jelly bag (cotton pillowcase or a double layer of cheesecloth) and squeeze out all of the liquid into a large pot. Add the peel and juice of 2 or 3 oranges and a lemon, ginger or any spice you prefer, to taste, and 3 pounds of sugar. Boil gently for 20 minutes. Remove from heat and cool to lukewarm. When lukewarm, put in a crock – either a gallon glass bottle with a water trap or an open crock with cheesecloth over the top. Then add the yeast – 1 tablespoon of baking yeast or purchase a special strain at a wine-making store. Ferment at 20-25°C for 6 days. Keep the temperature constant. If you have used a water trap, set the bottle in a dark place with a temperature of about 15°C for 2-3 months. With the crock method, after 6 days transfer the wine to a gallon glass bottle and plug with cotton wool. Store in a cool, dark place for 3 weeks. Decant the wine into bottles, and cap. Leave for about 5 months before drinking.

## Pin Cherry Wine

Put 12 cups of fruit in a large pot. Dissolve 6 cups of sugar in 2 quarts (8 cups) of boiling water. Add to cherries, stir well. When lukewarm, add yeast and ferment as for other fruit wines.

# Fruit Wine

The following recipe is a simple one for blueberries and raspberries. The amount of sugar can be modified for other fruit.

Add 1 gallon (16 cups) of boiling water to 8 cups of berries in a large pot, simmer for 20 minutes, and then boil for 5 minutes. Hang in a jelly bag (cotton pillowcase or a double layer of cheesecloth) over a pot and drain overnight. Measure the juice and add 4-6 cups of sugar for every gallon of juice. (Add 4 cups initially, stir, and taste.) Boil for 5 minutes and then cool. Pour into a container (see Dandelion Wine recipe) and add 3 cups of prunes for a dark juice like blueberry or sultana raisins for a lighter-coloured juice like raspberry (or use a yeast from a wine-making store). Let stand for 2 months at 15-18°C. Strain, bottle, and cork. The longer the wine ages in the bottle, the better it gets.

# Jelly

The type of fruit must be the first consideration in making jelly. Some fruit are quite juicy and so require less water to extract the juice. The extracted juice of all fruits should be boiled approximately 20 minutes before adding the sugar and for approximately 10 minutes after adding the sugar. It is best to use the amounts recommended – several small batches are better than one big batch.

Commercial fruit pectin, which is prepared from pectin-rich fruit like apples, is available in supermarkets. A recipe is given for squashberries. When making jelly from other fruit, alter the amount of sugar to suit the sweetness of the fruit.

## Squashberry Jelly

8 cups squashberries

7 cups water

Boil for 30 minutes. Hang in a jelly bag (cotton pillowcase or a double layer of cheesecloth) over a pot and drain overnight.

5 cups juice

7 cups sugar

Bring to a boil. Add 3 ounces of commercial pectin (or follow directions on the label), bring to a rolling boil (time from the point when you cannot stir enough to stop the tumbling boil) for 1 minute, and bottle.

# Rowan (Dogberry) Jelly
**(after Craig's recipe, 1965)**

2 pounds berries

Cold water as required

Rinse the berries and place in a preserving pot. Cover with cold water. Boil for 40 minutes. Strain through a jelly bag into a basin below, then measure the juice.

4 cups juice

7 cups sugar

1 bottle Certo

Return the juice to the pot. Add sugar. Bring to a full boil. Add Certo. Bring to a full boil again for 1 minute. Put in jelly jars and seal. Serve with roast game, chicken, or mutton, or with medium Cheddar cheese.

# Rose Hip Preserves

Rose hips are full of vitamin C in the outer red flesh and vitamin E in the seeds. Put 1 1/4 cups of rose hips (remove stalk and sepals), 3/4 cup of water, and 2 tablespoons of lemon juice in a blender and chop up hips. With blender operating, add 2 cups, or to taste, of sugar gradually. Blend 5 minutes. Transfer to a pot, simmer for 10-15 minutes, and strain. Bring to a rapid boil for 1 minute, stir in a commercial fruit pectin (follow label directions), and boil another minute. Bottle.

# Jam

| Fruit | Weight of fruit | Weight of sugar |
| --- | --- | --- |
| Chuckley Pear | 1 lb | 1 lb |
| Raspberry | 2 lbs | 1 lb |
| Blueberry | 2 lbs | 1 lb |
| Strawberry | 3 lbs | 2 lbs |
| Marshberry | 2 lbs | 2 lbs |
| Partridgeberry | 2 lbs | 2 lbs |

Measure the cups of fruit and place water, except for raspberries (use half Add the amount of sugar specified in the table and squash the fruit. Boil for 10 minutes, stirring occasionally. Add 2 tablespoons of lemon juice for each 2 pounds of blueberries after the final boil for extra flavour. Pour the jam into dry, sterilized jars and seal while hot.

# Partridgeberry Sauce

4 cups partridgeberries

3 cups water

2 cups sugar

Use a deep pot as this sauce boils up quickly, even when carefully watched. Put the berries and water in the pot and boil gently until all of the berries have popped (about 10 minutes). Stir in the sugar and gently boil for 5-10 minutes, stirring occasionally.

This recipe can also be used for marshberries and cranberries. For a thicker jam, use less water. Some variations include adding chipped apple with the berries or a dash of mace or nutmeg.

# Cranberry and Horseradish Relish

2 cups cranberries, chopped

½ cup horseradish

½ cup sugar

Juice of ½ a lemon

Mix and refrigerate for a few days so flavours can blend.

# Mint Jelly

Serve this with lamb, veal, or any meat from a young animal.

1 tbsp lemon juice

½ cup water

2 tbsp sugar

1 cup (packed) fresh mint

(young shoots)

Mix in a blender at high speed until liquefied. Add green food colouring, if desired. Prepare unflavoured gelatin (1/2 tbsp in 1/2 cup water) and add to mint. Allow to gel before serving.

# Cottage Pudding

This is a lovely way to use a few berries like blueberries, partridgeberries, crowberries, etc. You can throw a few in the batter and a few in the sauce.

2 cups sifted all-purpose flour
2 ½ tsp baking powder
¾ tsp salt
⅓ cup soft shortening
1 cup sugar
1 medium egg, beaten
1 cup minus 2 tbsp milk
1 tsp vanilla or 1 tsp grated orange or lemon peel

Preheat oven to 350°F. Put the shortening and sugar in a large mixing bowl and cream. Add the egg and beat until the mixture is light and fluffy (about 4 minutes). Sift flour, baking powder, and salt together. Add vanilla to milk (if fruit is added, reduce milk by ¼ cup). Blend the flour (by quarters) and milk (by thirds) mixtures alternately into the shortening mixture. Beat until just smooth. Fold in fruit or peel and turn into a well-greased 8 x 8 inch pan. Bake for about 45-50 minutes.

# Sauce

> 1 cup brown sugar
>
> 2 tbsp cornstarch
>
> 2⅓ cups water
>
> 1 tbsp butter
>
> Pinch of salt
>
> 1 tsp vanilla

Thoroughly mix brown sugar and cornstarch. Add water, butter, and salt and bring to a boil. Boil until the sauce thickens and then add vanilla and berries (optional). Boil 1 minute longer if berries have been added.

# Muffins

If you only have a few berries collected on an autumn stroll, make muffins.

> ¼ cup shortening
>
> ⅓ cup sugar
>
> 2 eggs, well-beaten
>
> 2 cups sifted all-purpose flour
>
> 4 tsp baking powder
>
> ¾ tsp salt
>
> ⅔ cup milk
>
> ⅔ cup blueberries, partridgeberries, etc.

Preheat oven to 400°F. Beat shortening until creamy; add sugar gradually, continuing to beat. Stir in eggs. Sift together baking powder, salt, and 1⅔ cups of flour. Add the flour mixture and milk alternately to the egg mixture; stir just enough to blend. Mix the berries with the remaining flour and gently stir into the batter. Spoon the batter into a greased muffin pan. Bake for 30 minutes or until brown. Makes 14 muffins.

# Pies

Use your favourite pastry recipe.

## Fruit Filling

The following recipe can be modified for other berries by varying the amount of sugar and/or omitting the lemon juice.

## Blueberry Pie Filling

3 cups blueberries

⅔ - 1 cup sugar

1 tbsp butter

3½ tbsp cornstarch (optional)

1½ tbsp lemon juice (optional)

Sprinkle the other ingredients over the berries in a bowl and mix gently. Pour into an unbaked pie crust and dot with butter. Seal with a top crust and bake at

350°F for 1 hour or until browned.

# Partridgeberry Tarts

This is a lovely recipe from Notre Dame Bay.

4 cups all-purpose flour
2½ tsp cloves
2½ tsp cinnamon
1 tsp ginger
1 cup margarine
¾ cup molasses
2 tsp baking soda in ¼ cup of tea
Partridgeberry jam

Preheat oven to 425°F. Sift together the dry ingredients. Cream margarine, beat in molasses, and mix in baking soda mixture. Stir in dry ingredients. Roll and line tart tins. Fill with partridgeberry jam. Bake 10-12 minutes.

# Quick Bread

This is one of many recipes that can be used with native fruit.

1½ cups pecans (or walnuts), divided (see below)
1½ cups coarsely chopped cranberries (or whole marshberries)
1¼ cups sugar
3 cups all-purpose flour
4½ tsp baking powder
½ tsp salt
2 tsp grated lemon rind
½ cup vegetable shortening
2 eggs, well-beaten
1 cup milk

Preheat oven to 350°F. Grease a loaf pan and sprinkle ½ cup of finely chopped pecans evenly over the bottom. Mix cranberries and ¼ cup sugar in a bowl. In a large bowl, sift flour, remaining sugar (1 cup), baking powder, and salt. Cut in shortening with a pastry blender until mixture resembles cornmeal. Stir in 1 cup of coarsely chopped pecans and lemon rind. Stir milk into eggs, add all at once to

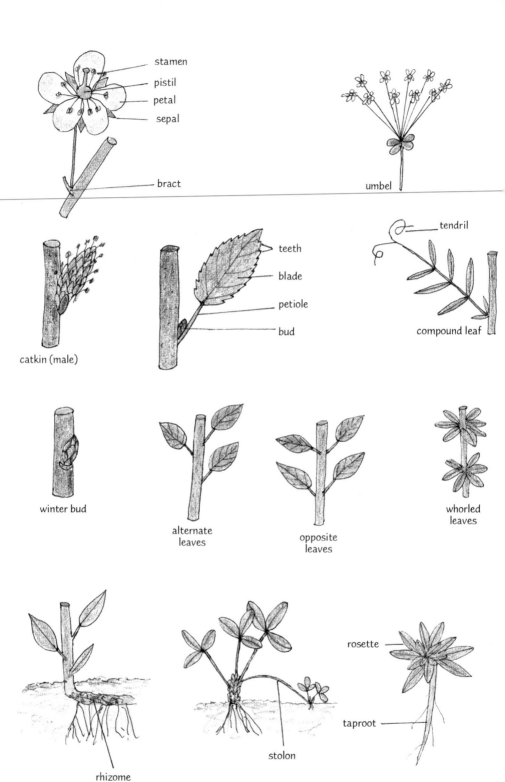

stamen
pistil
petal
sepal

bract

umbel

teeth
blade
petiole
bud

catkin (male)

tendril

compound leaf

winter bud

alternate
leaves

opposite
leaves

whorled
leaves

rhizome

stolon

rosette

taproot

# Glossary

## *Flowers*

Flowers have four sets of structures. Starting from the outside of a flower:

**sepals** are usually green and leaflike in texture.

**petals** are colourful and more delicate in nature: in many flowers, they attract pollinators.

**stamens** are the male structures which produce pollen.

**pistils** are the female structures which produce seeds.

**umbel** – a flat-topped grouping of flowers where the flower stalks are attached to a common point on the tip of a stalk.

**catkin** – a highly modified grouping of flowers which are pine-cone-like with scaly bracts. There are separate male and female catkins.

## *Leaves*

Leaves have a leaf stalk (**petiole**) and a **blade**.

A **simple leaf** has a single blade, which may be lobed (e.g., maple).

A **compound leaf** has a number of leaflets.

The margins of a leaf or leaflet can be smooth or jagged with **teeth**.

## *Arrangement of leaves*

**rosette** – leaves radiate from a centre, usually on the top of the root at soil level.

Leaves can be attached to the stem

singly = **alternate**,

in pairs = **opposite**,

or in a circle at a point on the stem = **whorled**.

**winter buds** are found at the base of leaves on trees and shrubs. They are usually oval in shape and have hard, overlapping scales which protect the dormant branches through the winter.

**bracts** are small leaves found in flower clusters (= inflorescences).

**fronds** – a special term for the leaves of ferns.

**tendril** – a threadlike modification of a leaf or stem which coils around other plants for support.

# Roots

Roots take two main forms:

**fibrous** root systems consist of a mass of fine branching roots.

**taproots** have thick main roots with fine side roots (e.g., carrot, dandelion).

# Stems

**rhizome** – many herbs have a thickened stem which grows at or just below the soil surface.

**stolon** – an elongated branch which bends to the soil, roots, and grows a new plant at that point.

# Geography

**boreal** – the region of the Northern Hemisphere covered with boreal or northern forest which is dominated by conifers (mainly spruce and fir).

**circumboreal** – a plant geography term which indicates that a species occurs in the boreal regions of North America, Europe, and Asia.

**Eurasia** – a plant geography term which indicates that a species occurs across Europe and Asia.

# APPENDIX – Plant Names

Dennstaedtiaceae – Wood Fern Family
> *Pteridium aquilinum* (L.) Kuhn – Bracken

Dryopteridaceae – Wood Fern Family
> *Matteuccia struthiopteris* (L.) Todaro – Ostrich Fern

Cupressaceae – Cypress Family
> *Juniperus communis* L. – Common Juniper
> *Juniperus horizontalis* Moench – Trailing Juniper

Pinaceae – Pine Family
> *Picea glauca* (Moench) Voss – White Spruce
> *Picea mariana* (Mill.) BSP – Black Spruce

Taxaceae – Yew Family
> *Taxus canadensis* Marsh. – Canada Yew

Adoxaceae – Viburnum Family
> *Sambucus racemosa* L. subsp. *pubens* (Michx.) House
> > – Red Elderberry
> *Viburnum edule* (Michx.) Raf. – Squashberry
> *Viburnum nudum* L. var. *cassinoides* (L.) Torr. & Gray
> > – Northern Wild Raisin
> *Viburnum opulus* L. var. *americanum* Ait.
> > (syn. *Viburnum trilobum*) – Highbush Cranberry

Amaranthaceae – Amaranth Family
> *Atriplex glabriuscula* Edmonston – Smooth Orach
> *Atriplex patula* L. – Orach
> *Chenopodium album* L. – Lamb's Quarter

Apiaceae – Parsley Family

    *Angelica atropurpurea* L. – Angelica

    *Ligusticum scothicum* L. – Scotch Lovage

Asteraceae – Daisy Family

    *Taraxacum officinale* Weber – Dandelion

    ~~*Tussilago farfara* L. – Coltsfoot~~

Betulaceae – Birch Family

    *Alnus incana* (L.) Moench subsp. *rugosa* (DuRoi) Clausen

        (syn. *A. rugosa*) – Speckled Alder

    *Alnus viridus* (Chaix) DC subsp. *crispa* (Dryand.) Turrill

        (syn. *A. crispa*) – Mountain Alder

    *Betula alleghaniensis* Britt. – Yellow Birch

    *Betula papyrifera* Marsh. – White Birch

    *Corylus cornuta* Marsh. – Beaked Hazelnut

Brassicaceae – Mustard Family

    *Barbarea vulgaris* Ait. – Winter Cress

    *Cakile edentula* (Bigel.) Hook. – Sea Rocket

    *Capsella bursa-pastoris* (L.) Medic. – Shepherd's Purse

    *Cochlearia officinalis* L. – Scurvygrass

    *Cochlearia tridactylites* Banks – Three-fingered Scurvygrass

    *Nasturtium microphyllum* Boenn. – Watercress

        (syn. *N. officinale*)

    *Thlaspi arvense* L. – Pennycress

Caprifoliaceae – Honeysuckle Family

    *Lonicera villosa* (Michx.) R. & S. – Northern Fly Honeysuckle

Caryophyllaceae – Pink Family

    *Stellaria media* (L.) Villars – Common Chickweed

Cornaceae – Dogwood Family
    *Cornus canadensis* L. – Crackerberry
    *Cornus suecica* L. – Swedish Bunchberry

Crassulaceae – Stonecrop Family
    *Rhodiola rosea* L. – Roseroot

Elaeagnaceae – Oleaster Family
    *Shepherdia canadensis* (L.) Nutt. – Soapberry

Ericaceae – Heath Family
    *Andromeda glaucophylla* Link – Bog Rosemary
    *Arctostaphylos uva-ursi* (L.) Spreng. – Bearberry
    *Chamaedaphne calyculata* (L.) Moench – Leatherleaf
    *Empetrum atropurpureum* Fern. & Wieg. – Purple Crowberry
    *Empetrum eamesii* Fern. & Wieg. – Pink Crowberry
    *Empetrum nigrum* L. – Crowberry
    *Epigaea repens* L. – Trailing Arbutus
    *Gaultheria hispidula* (L.) Muhl. – Creeping Snowberry
    *Gaylussacia baccata* (Wangenh.) Koch – Black Huckleberry
    *Gaylussacia dumosa* (Andr.) T. & G. – Dwarf Huckleberry
    *Rhododendron groenlandicum* (Oeder) K. & J. – Labrador Tea
        (syn. *Ledum groenlandicum*)
    *Rhododendron lapponicum* (L.) Wahlenb. – Lapland Rosebay
    *Vaccinium angustifolium* Ait. – Blueberry
    *Vaccinium macrocarpon* Ait. – Cranberry
    *Vaccinium oxycoccus* L. – Marshberry
    *Vaccinium uliginosum* L. – Alpine Bilberry
    *Vaccinium vitis-idaea* L. – Partridgeberry

Fabaceae – Pea Family
    *Lathyrus japonicus* Willd. – Beach Pea
    *Trifolium hybridum* L. – Alsike Clover

*Trifolium pratense* L. – Red Clover
*Trifolium repens* L. – White Clover

Grossulariaceae – Currant Family
*Ribes glandulosum* Grauer – Skunk Currant
*Ribes hirtellum* Michx. – Smooth Gooseberry
~~*Ribes lacustre* (Pers.) Poir. – Bristly Black Currant~~
*Ribes triste* Pallas – Swamp Red Currant

Lamiaceae – Mint Family
*Mentha arvensis* L. – Wild Mint
*Mentha cardiaca* Baker – Heart Mint

Liliaceae – Lily Family
*Clintonia borealis* (Ait.) Raf. – Corn Lily

Malvaceae – Mallow Family
*Malva moschata* L. – Muskmallow

Myricaceae – Wax-Myrtle Family
*Myrica gale* L. – Sweet Gale

Nymphaeaceae – Waterlily Family
*Nuphar variegata* Durand – Bullhead Lily
*Nymphaea odorata* Dryand. – Fragrant Waterlily

Onagraceae – Evening Primrose Family
*Chamerion angustifolium* (L.) Holub – Fireweed
(syn. *Epilobium angustifolium*)

Osmundaceae – Flowering Fern Family
*Osmunda cinnamomea* L. – Cinnamon Fern

Plantaginaceae – Figwort Family
 *Plantago maritima* L. – Seaside Plantain

Poaceae – Grass Family
 *Ammophila brevilingulata* Fern. – American Beachgrass
 *Leymus arenarius* (L.) Hochst. – Strand Wheat

Polygonaceae – Buckwheat Family
 *Fallopia japonica* (Houtt.) Ronse-Decr. – Japanese Knotweed
 *Fallopia sachalinensis* (Schmidt) Ronse-Decr. – Giant Knotweed
 *Rumex acetosa* L. – Garden Sorrel
 *Rumex acetosella* L. – Sheep Sorrel
 *Rumex crispus* L. – Curled Dock

Ranunculaceae – Buttercup Family
 *Caltha palustris* L. – Marsh Marigold

Rosaceae – Rose Family
 *Amelanchier bartramiana* (Tausch) Roemer – Bartram's Chuckley Pear
 *Amelanchier canadensis* (L.) Medic. – Canadian Chuckley Pear
 *Amelanchier interior* E.L. Nielsen – Wiegand's Chuckley Pear
 *Amelanchier laevis* Wiegand – Smooth Chuckley Pear
 *Amelanchier sanguinea* (Pursh) DC – Fernald's Chuckley Pear
 *Amelanchier stolonifera* Wiegand – Running Chuckley Pear
 *Fragaria vesca* L. – Woodland Strawberry
 *Fragaria virginiana* Mill. – Wild Strawberry
 *Photinia floribunda* (Lindl.) Robertson & Phipps –
   Purple Chokeberry
 *Photinia melanocarpa* (Michx.) Robertson & Phipps –
   Black Chokeberry
 *Prunus pensylvanica* L.f. – Pin Cherry
 *Prunus virginiana* L. – Chokecherry

*Rosa nitida* Willd. – Northeastern Rose

*Rosa virginiana* Mill. – Virginian Rose

*Rubus canadensis* L. – Canadian Blackberry

*Rubus chamaemorus* L. – Bakeapple

*Rubus hispidus* L. – Bristly Blackberry

*Rubus idaeus* L. – Raspberry

~~*Rubus pensilvanicus* Roir. – Pennsylvanian Blackberry~~

*Rubus pubescens* Raf. – Hairy Plumboy

*Sorbus americana* Marsh. – American Dogberry

*Sorbus aucuparia* L. – European Mountain Ash

*Sorbus decora* (Sarg.) C.K. Schneid. – Northern Dogberry

*Sorbus groenlandica* (C.K. Schneid.) A & D Love –
Greenland Mountain Ash

Rubiaceae – Madder Family

*Galium* spp. – Bedstraw or Cleavers (23 species in Canada; 11 species in Atlantic Canada)

Typhaceae – Cattail Family

*Typha latifolia* L. – Cattail

Urticaceae – Nettle Family

*Urtica dioica* L. subsp. *gracilis* (Ait.) Solander – Stinging Nettle

*Urtica urens* L. – Burning Nettle

Violaceae – Violet Family

*Viola cucullata* Ait. – Blue Marsh Violet (13 species in Atlantic Canada)

Note

syn. – synonyms are previous names for species. Some are still used in reference books such as field guides.

# References

Craig, Elizabeth. 1965. *What's Cooking in Scotland*. Edinburgh: Oliver & Boyd.

Erichsen-Brown, Charlotte. 1979. *Use of Plants for the Past 500 Years*. Aurora, ON: Breezy Creek Press.

Fernald, M.L. and A.C. Kinsey. 1943. *Edible Wild Plants of Eastern North America*. NY: Harper & Row.

Gerard, John. 1633. *The Herbal*. NY: Dover Publications.

Gibbons, Euell. 1962. *Stalking the Wild Asparagus*. NY: David McKay Co.

Gibbons, Euell. 1964. *Stalking the Blue-Eyed Scallop*. NY: David McKay Co.

Gordon, L.D. et al. (ed.). 1986. *Magic and Medicine of Plants*. Montreal: Reader's Digest Association.

Obed, Ellen Bryan. 2008. *Partridgeberry, Redberry, Lingonberry, Too*. Portugal Cove-St. Philip's: Boulder Publications.

Scott, Peter J. 1979. *Some Edible Fruit and Herbs of Newfoundland*. St. John's: Breakwater Books.

Wood, William. 1634. *New England Prospect*. London.

The following were quoted in Erichsen-Brown (1979):

Bigelow, Jacob. 1817-1820. *American Medical Botany*. Boston: Cummings & Hilliard.

Bye, Robert A., Jr. 1970. *The Ethnobotany and Economic Botany of Onandaga County, NY.* Unpublished thesis, Harvard University.

Delamare, E., F. Renauld, and J. Cardot. 1888. *Florula de l'ile Miquelon.* Lyon, France: Assoc. Typographique.

Gosse, P.H. 1840. *The Canadian Naturalist. The Natural History of Lower Canada.* London: John van Voorst.

Isham, James. 1743. *Observations on Hudson's Bay.* London: Rich & Johnson.

Kalm, Peter. 1748-1751. *Travels in North America.* NY: Wilson-Erickson.

Mason, John. 1620. *A Briefe Discourse on the New-found-land.* Edinburgh: Andro Hart.

Medsger, O.P. 1939. *Edible Wild Plants.* NY: MacMillan.

Rafinesque, C.S. 1830. *Medical Flora.* Philadelphia: Samuel C. Atkinson.

Richardson, James Henry. 1829-1905. *Reminiscences.* Toronto: University of Toronto Press.

Sagard, Theodat Gabriel. 1624. *Le gran voyages du pays des Hurons.* NY: Greenwood (translation).

Smith, Capt. John. 1612. *Description of Virginia and proceedings of the colony.* Oxford.

Sturtevant, E. Lewis. 1919. *Sturtevant's Notes on Edible Plants.* Albany: Lyon.

Traill, Catherine Parr. 1838. *The Backwoods of Canada.* London: C. Knight.

# Photo Credits

**Todd Boland**: Angelica, Bakeapple, Banks and Shores, Beach Pea, Blue Marsh Violet, Blueberry, Canadian Blackberry, Cattails, Chuckley Pear, Clearings, Cleavers, Corn Lily, Crackerberry, Cranberry, Crowberry, Curled Dock, Disturbed Areas, Dogberry, Fiddleheads, Fireweed, Highbush Cranberry inset, Marshberry, Mountain Alder, Northeastern Rose, Northern Fly Honeysuckle, Northern Wild Raisin inset, Orach, Peatlands, Pin Cherry, Raspberry, Red Clover, Roseroot, Scotch Lovage, Scurvygrass, Sea Rocket, Seaside, Seaside Plaintain, Skunk Currant, Soapberry, Spruce, Squashberry, Stinging Nettle, Strand Wheat, Wild Birch, Wild Strawberry inset, Winter Cress

**Andy Fyon:** Canada Yew inset, Chickweed, Chokecherry, Creeping Snowberry, Hairy Plumboy, Japanese Knotweed, Labrador Tea, Mint, Northern Wild Raisin, Skunk Currant inset, Sorrel, Trailing Arbutus, Watercress

**Louis-M. Landry:** Beaked Hazelnut, Canada Yew, Chokeberry, Common Juniper, Pennycress

**Peter Scott:** Drawing (p. 158), Forest Floor, Heaths, Waterlily

**John Maunder:** Creeping Snowberry inset, Huckleberry inset

**Robert E. Young:** Dandelion

**Gerald Tang:** Lamb's Quarter

**PlantWatch Canada:** Wild Strawberry

**U.S. Department of Agriculture:** Squashberry inset

**Kohrs Images:** Highbush Cranberry

**cretanfllora.com:** Shepherd's Purse

**istockphoto.com:** Marish Marigold, Recipes

# Acknowledgements

The natural beauty of our province has been a constant inspiration for the books that I have written.

I would like to thank my friends and colleagues, Dr. Jean R. Finney-Crawley and Dr. A.J. Shawyer, for their help in preparing this volume. They read the manuscript and made many valuable comments. I am also grateful to Iona Bulgin for her painstaking editing.